JOBS KIDS WANT

WHAT'S IT REALLY LIKE TO BE A VETERINARIAN?

CHRISTINE HONDERS

PowerKiDS press.
New York

Published in 2020 by The Rosen Publishing Group, Inc.
29 East 21st Street, New York, NY 10010

First Edition

Editor: Greg Roza
Book Design: Michael Flynn

Photo Credits: Cover, p. 1 LWA/DigitalVision/Getty Images; pp. 4, 6, 8, 10, 12, 14, 16, 18, 20, 22 (background) Apostrophe/Shutterstock.com; p. 5 wavebreakmedia/Shutterstock.com; p. 7 https://commons.wikimedia.org/wiki/Category:History_of_veterinary_medicine#/media/File:Apothecary_giving_medicine_to_dog,_Augsburg_school,_18th_c._Wellcome_L0019359.jpg; p. 9 Fuse/Corbis/Getty Images; p. 11 Tyler Olson/Shutterstock.com; p. 13 Orlando Sierra/AFP/Getty Images; p. 15 Bruce Weber/Shutterstock.com; p. 17 VGstockstudio/Shutterstock.com; p. 19 hedgehog94/Shutterstock.com; p. 21 Robert Daly/Caiaimage/Getty Images; p. 22 Terayut/Shutterstock.com.

Cataloging-in-Publication Data

Names: Honders, Christine.
Title: What's it really like to be a veterinarian? / Christine Honders.
Description: New York : PowerKids Press, 2020. | Series: Jobs kids want | Includes glossary and index.
Identifiers: ISBN 9781725300286 (pbk.) | ISBN 9781725300309 (library bound) | ISBN 9781725300293 (6 pack)
Subjects: LCSH: Veterinarians–Juvenile literature. | Veterinary medicine–Vocational guidance–Juvenile literature.
Classification: LCC SF756.H66 2020 | DDC 636.089'069–dc23
Manufactured in the United States of America

CPSIA Compliance Information: Batch #CSPK19. For Further Information contact Rosen Publishing, New York, New York at 1-800-237-9932.

CONTENTS

Animal Doctors

Pets are part of our families. When they get sick, we take them to a veterinarian. Veterinarians are doctors for animals. Veterinarians do more than give shots to cats and dogs. They keep both animals and people from getting sick.

Veterinarians in History

The first veterinarians may have worked around four thousand years ago! In 1700, a **disease** wiped out almost all the cattle in Europe. Soon after, people began studying how to keep animals from getting sick. That is when modern veterinarian practice began.

What Veterinarians Do

Veterinarians figure out why animals are sick and how to take care of them. They **examine** animals and test their blood. They give them **medicine** to make them better. Veterinarians give animals shots to keep them healthy. Some veterinarians do **surgery** on animals.

At the Office

Many veterinarians work in offices. They do checkups and care for sick pets. They tell cat and dog owners how to keep their pets free from fleas. Some veterinarians work in animal hospitals. They care for very sick or **injured** animals.

House Calls

Some veterinarians take care of farm animals. They make sure the animals are disease-free and don't get each other sick. They even help animals give birth to their babies! Veterinarians also work to keep zoo animals healthy. They help keep **endangered** animals safe.

Keeping Us All Healthy

Veterinarians also keep people safe. They fight diseases that animals can give to people. They study new medicines that cure these diseases. Some veterinarians go to places where meat is packed to be sold. They examine the animals to make sure the meat is safe to eat.

Understanding Animals

Many animals are scared when they go to the veterinarian. Veterinarians know how to calm them down. Animals can't talk, so veterinarians watch them carefully. They find out from the owner how they've been acting at home. This helps them figure out what's wrong.

Day or Night, Rain or Shine

Like people doctors, veterinarians have long days. They work evenings and weekends. They run out in the middle of the night if there's an emergency. Veterinarians working for farms and zoos are often outside. They work in any weather, all year long.

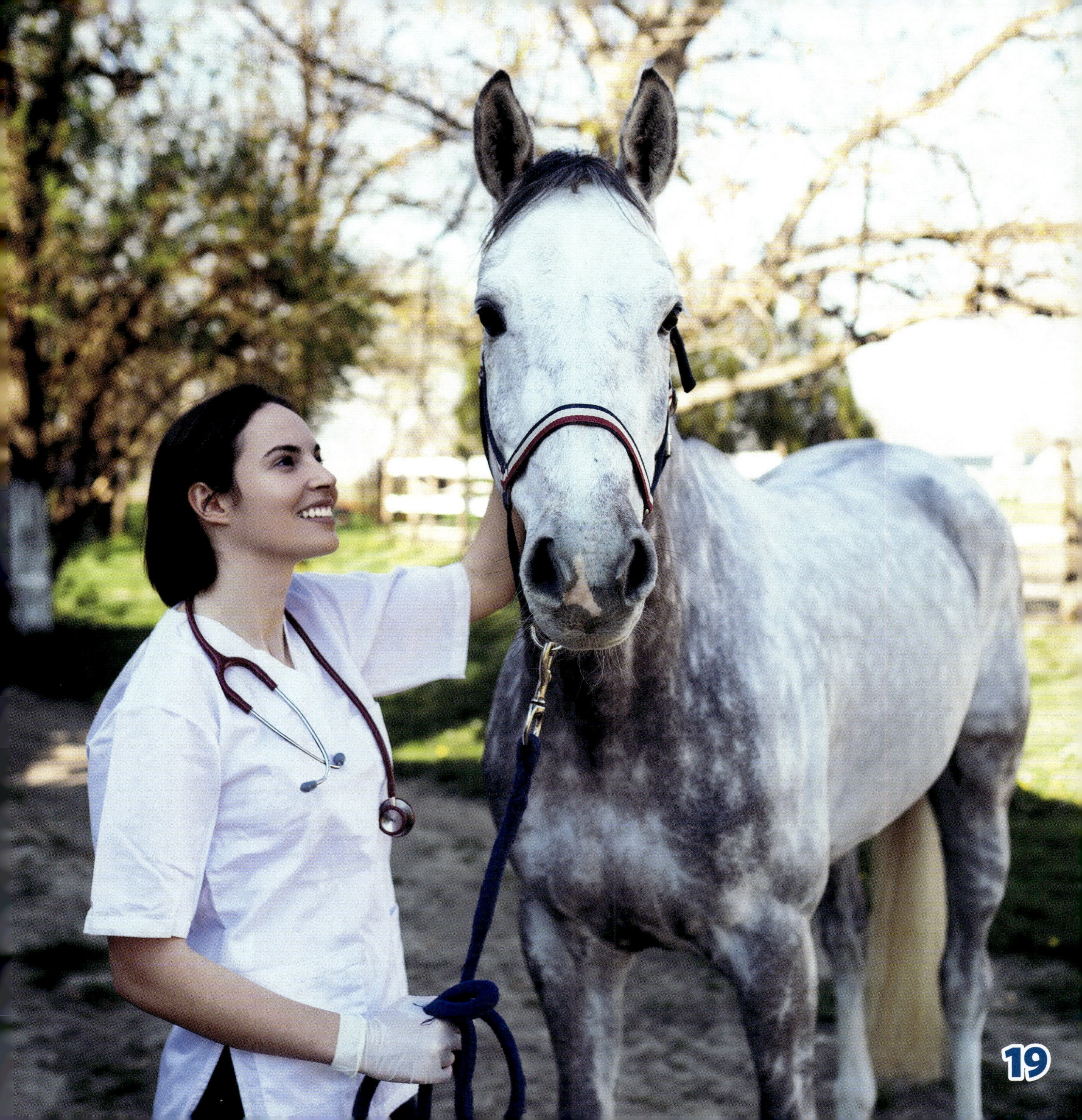

Becoming a Veterinarian

After high school, students who want to be veterinarians go to college. They study **biology** or another science. Then they take a test to get into veterinary school. Veterinary school lasts four years. Students then take one final test. If they pass, they're veterinarians!

Good Vets Love Pets

Most veterinarians say they're very happy in their jobs. They love animals and want to keep them healthy and safe. They like making pet owners feel better. Veterinarians give a voice to our pets. But they make a difference in all our lives!

GLOSSARY

biology: The study of all living things.

disease: Illness.

endangered: At risk of disappearing from the planet for good.

examine: To look at very carefully.

injured: Hurt or in pain.

medicine: A drug someone takes to make them feel better.

surgery: Medical treatment that involves cutting into the body.

INDEX

WEBSITES

Due to the changing nature of Internet links, PowerKids Press has developed an online list of websites related to the subject of this book. This site is updated regularly. Please use this link to access the list: www.powerkidslinks.com/JKW/veterinarian